The Runaway

by
Billie West Paine

"The Runaway" © 2005
All Rights Reserved.

No portion of this publication may be reproduced in any manner, without the prior permission of the publisher.

Cover design and text formatting by Beth E. Miller

ISBN: 0-932397-22-0 (paperback)

International Biblical Resources

Dedicated to my children

Rod, Pam, Russ, and Rex, who became
the shining stars in the life of the main
"character" in this book,
their grandfather!

Acknowledgments

<u>Olga Money</u> - my older sister, who remembered things I didn't know.

<u>Rod</u> - the oldest grandchild who would sit for hours listening to his grandfather's stories (and remembering them!)

<u>Rex</u> - who took me to Scotland to do research.

<u>Charlsie</u> - Captain Dick's daughter who gave me information I had never heard.

<u>Beth Miller</u> - without her help there would be no book.

Preface

The Scots and the Irish are well known for their storytelling skills. My dad, William George West, which may or may not be his real name, was no exception. In any group of people, on any occasion, regardless of the topic of conversation, he would soon be the one speaking, and the group sitting with rapt attention. It wasn't that he forced himself on people. He would be sitting quietly and others drew him into the conversation. They became so fascinated by the things he had to say that he ultimately became the focus of the group. The Scottish brogue, the raspy, yet lilting voice telling of his sometimes unbelievable adventures as a man of the sea, etched this man forever in the listener's memory!

Foreword

I am one of many thousands of Americans who, in our old age, discovered we have Jewish ancestors. My older sister Olga and I grew up in an "emigrant" middle class neighborhood on Galveston Island just off the coast of Texas in the Gulf of Mexico. I never felt different from any other people, for Galveston was full of "foreigners." As a child I thought all cities were like that. On our little street, on our block we had: the McGuires from Ireland, the Rollisches from Yugoslavia, the Shumans and the Shadts from Germany, Gus the Swede and the Magliolos from Italy! The corner grocery, Miro's Market, where we shopped was owned and operated by Mr. and Mrs. Miro from Greece. My first swimming instructor was a beach lifeguard called Nick "the Greek." I can still hear his voice as he constantly reminded me to "move your 'hanses' slow and your 'feetses' fast!" He was a great teacher. My dad had a heavy Scottish brogue and often I would have to "translate" his conversation for my friends. I thought nothing of it.

Because of the times and circumstances of his early life, my dad "reinvented" himself at an early age. This probably included a name change. My only regret is his mysterious past has left us with no known relatives on his side of the family. Just as the waves constantly sliding upon the shore wipe away sandcastles, so it seems the beaches of my father's past have been swept clean both by time and intent. He completely reinvented himself!

Contents

Dedication
Acknowledgments
Preface
Forward

I.	The Decision	1
II.	The First Voyage	9
III.	Death of Queen Victoria	17
IV.	The First Shipwreck	23
V.	The Return	31
VI.	The School	43
VII.	Another Ship Lost, Another Long Walk	49
VIII.	The African Adventure	51
IX.	Two Wars	63
X.	Smitten With Malaria And LOVE! ...	83
XI.	Farewell To Captain MacDonald	95
XII.	The Orphan	101
XIII.	Back To School	107
XIV.	A Life Changing Experience	113

"The Box" 129
Monahans Newspaper Reprint 133
Epilogue 137
Endnotes 141

Chapter 1

Chapter 1

The Decision

His heart was pounding as he stood in front of the burly sailor who was sitting behind a rough-hewn table. He began to wonder if the sailor would brush him aside with a coarse laugh or take to the magistrate to be held as a runaway. He was small for a twelve year old, though muscular and strong. His uncle had seen to that! He had used every muscle in his small frame working out in the cold wind in the Highlands of Scotland. He shifted his weight nervously as the sailor looked up from a handful of yellowed papers.

"What 'ave we here?" said the sailor, peering above his glasses that had slipped down on his nose. His face had a rough and ruddy weathered look, yet his piercing blue eyes were kindly as they passed over

The Decision

the boy, sizing him up. "What brings a lad like 'yer'self out here to the docks on such a cold morning?" the sailor said in his thick Scottish brogue.

West struggled to keep his gaze steady but was able to look the man straight in his eyes as he answered, "I'm hopin' you're in need of a cabin boy, for I've heard talk that you're soon to put out to sea." He hastened to add, "I'm familiar with ships and I'm strong. I can climb the masts, I know how to tie off the sails and I can even cook!" The sailor chuckled as he said, "Well, son, I don't see as how I could afford ***not*** to take you on with all 'yer' skills. And what would you be callin' yerself, so I can be writing yer name down?" West swallowed and hoped Captain John MacDonald wouldn't notice his nervousness as he answered, *"William George West,* sir."

This event began a friendship between "Big John MacDonald" and West that would last for years as they took many voyages together.

West was so excited that he wanted to jump up and click his heels, but he restrained himself and simply answered with, "Thank you sir and you'll not be regrettin' hiring me on for I'll do my best to make myself useful to ye!" Captain MacDonald stood up

The Decision

and West was surprised at the size of the man, yet the Captain said in a gentle voice, "If ye be needin' a place to sleep tonight you can sleep aboard the Arabella. We'll be finished loading our cargo today and cast off at first light on the morrow." He glanced down at the boy's small canvas bag and wondered what his story was – that one so young should be hustling a job on his own.

When MacDonald saw the relief on the boy's face he motioned to one of his hands nearby and said "Simpson, show our new cabin boy to the small room next to mine and see that he gets some breakfast. We can't have a man workin' on an empty stomach! West, when you get yerself settled in, go to the galley and after you have yer breakfast, help Antonio, the cook, load the stores." West stuck out his hand and as they shook hands, Big John was surprised at the strength and hardness of the hand, for such a young lad.

West immediately liked Antonio who prepared a generous breakfast for him – ham and eggs, big slices of bread with butter and hot coffee. Antonio was a handsome Italian man with smooth olive skin and large brown eyes framed with long black lashes.

The Decision

His build was slight though muscular and several inches taller than West. West thought his looks and winning smile must have left many broken hearts behind! Antonio talked the whole time West was eating, as well as, while they were stocking the shelves with the food stores for their trip. He told West all about his family in Italy. West got the feeling that Antonio was lonesome for someone to talk with or maybe he was just missing his family. He didn't ask West any questions, so he just listened. West listened closely as most of his conversation was in Italian.

The day passed quickly as West was "making himself useful" stowing the smaller cargo items in wooden chests and fetching tools and ropes for anyone who asked. He was up and down the ladder into the hold, lashing down the cargo with ropes so it wouldn't shift when they got underway. He was even up the masts as final repairs were made.

That night as he laid on his bunk he thought he must be dreaming! He had a job on a good ship with a good captain and best of all the ship would be sailing right away! No one would think to look for him aboard ship. He had left his uncle's place in the

The Decision

Highlands around midnight when he was sure his uncle was fast asleep. He had walked swiftly, even running part of the way. He wanted to get to Aberdeen in time to locate the docks. He had been here before, but never on his own. Those were the longest 30 miles he had ever walked!

West undressed quickly and crawled into his bunk. It felt wonderful to stretch out between the soft blankets! Every muscle in his body was aching. In all his excitement and longing to get some sleep, he had forgotten something. He got up from his bunk and lit the candle on the small crate that served as a bedside table. The canvas bag he had brought on board was on the floor nearby. It contained only a few items: a change of clothing, a slicker, a warm jumper and two books. He slipped the books out carefully. One was a book of poetry by Rudyard Kipling and the other was a Bible. As  he opened the Bible a picture fell out. He picked up the picture and looked at it wistfully for a minute or two, then carefully placed it back between the pages. It was a picture of his mother.

He crawled back into his bunk and held the Bible close to the candle so he could read. After a little reading, he gently laid his Bible on the crate and blew out the candle. He said a prayer thanking God for his good fortune.

Sleep came quickly after all the excitement and exhaustion of the day and the long walk the night before. Would anyone be looking for him? Or would his uncle just say "good riddance!"

Chapter 2

Chapter 2

The First Voyage

West was aroused early by noises coming from the galley. He was a very light sleeper. This characteristic would stay with him the rest of his life. In moments he had bolted from his bunk, splashed his face with cold water from the stainless steel basin located on the bedside crate, dressed and pulled on his warm jumper. Though it was August, there was a chill in the early morning air.

After lending a hand to Antonio, West went up on deck just in time to see the sun creeping over the horizon. He noticed that the North Star was still visible. Simpson and the other hands were loosening the sails and preparing to cast off. The cold wind of the night before had settled itself to a more gentle

The First Voyage

breeze. A light misty rain began to fall as the Arabella slipped away from the dock. The clouds that brought the rain were beginning to obscure the rising sun. West felt a sense of excitement and relief as they were getting underway.

West's duties aboard ship included keeping the galley floors mopped, making sure the Captain's quarters were clean and tidy. He was also responsible for keeping the ship's lanterns full of oil and the globes clean as well as making any minor repairs requested by Captain MacDonald or Simpson. He soon became quite skilled at repairing anything mechanical. He was frequently called upon  to polish the brass port holes and other brass or copper fittings that made the Arabella cause people to give her a "second look" when she was in port!

The Arabella was a sturdy three masted schooner. She was larger than most of the ships West had been on since nearly all of them were fishing

The First Voyage

vessels. He noticed how clean and well kept she was. West loved the sea and had gone on board ships at every opportunity. Coming from a seafaring clan, it seemed natural that he was launching out on his own aboard ship. He tried not to think about the events that brought him to this day!

Soon the Aberdeen docks and the Scottish coast were barely visible, clothed in dense fog. The year was 1900 and many historical events were taking place around the world. The Boer War in Africa was getting the attention of the British World. Two of West's older brothers had signed up the year before. England sent 250,000 men to South Africa. This was the largest army ever put in the field in the course of British history.[1] The Orange Free State united with the Transvaal against Great Britain. This proved to be a bitter and expensive war. West heard all about this when four men came to see about buying some of his uncle's sheep. Their sons had written home telling about several of the battles that involved the men in their company.

Unknown to the crew of the Arabella, another monumental event was taking place across the Atlantic. On September 8, 1900 the most devastating

The First Voyage

storm ever to hit North America, was pounding the historic city of Galveston, Texas, an island in the Gulf of Mexico just off the Texas coast. A wall of water taller than a two story house along with very high winds, swept over the island, demolishing everything in her path. Between 6,000 and 8,000 people lost their lives in one day. Galveston had been called "the New York of the Gulf Coast." It boasted such modern amenities as electric lights, streetcars, a telephone

The First Voyage

system and an opera house. Galveston was Texas' pre-eminent city in the year of 1900, she had a thriving port, industry, commerce, and culture. In earlier days Galveston had been the headquarters for the infamous pirate, Jean Lafayette. West had no way of knowing that this city would play a huge role in his future life![2]

Chapter 3

Chapter 3

Death of Queen Victoria

The Arabella had been to sea a scant six months stopping off at ports in Stockholm, Copenhagen, and Amsterdam, unloading and loading cargo, when they got the news that Queen Victoria had died on January 21, 1901. West had just turned 13 on January 7th. The men got the news of the Queen's death from the crew of another ship called "The Star" who had arrived in Amsterdam from England the day before the Arabella. They met up with the men from the Star in one of the pubs and West noticed that the crew of the Arabella seemed to be acquainted with them. Most of the Star's crew were Dutch. When Simpson introduced West to the Star's crew, he was surprised that West communicated with them in the Dutch language. West had been to Amsterdam several times before!

Another news item the crew learned was concerning the new King of Great Britain. They were told that the Prince of Wales was crowned King Edward VII. He was well liked and had a unique ability as a conciliator. He was able to complete the work of uniting the British Colonies, which was begun by Queen Victoria. Unfortunately, he was more successful on foreign soil (the colonies) than he was domestically!

The British Empire, amounted to one fifth of the land surface of the globe![3] This colony expansion took place between 1837 and 1910, mostly during the reign of Queen Victoria. In 1846, trade restrictions that had prevented trade between the colonies and the rest of the world, were lifted. This made trade extremely lucrative. Prior to the repealing of this chauvinistic "Jingo" law, the colonies of England could not trade with anyone outside the British

Commonwealth. The expression "By Jingo" originated from this law.

During these first six months, West was beginning to find his place among the crew. His favorites besides Captain MacDonald, were Antonio, the Italian cook, Gus, the Swede, Alex, the Irish and two brothers from Scotland, Jack, and Harry. Of course, Simpson who had been so helpful that first day, was his special friend. Simpson, the first mate, had sailed with Captain MacDonald for about ten years. He was an expert in jujitsu. He began to teach West some of the aggressive tactics and defensive moves. He also was skilled in boxing and instructed West in this as well. West learned quickly as he had good coordination and athletic skills. When their duties were done, it became a routine to meet on deck and practice for an hour or two, weather permitting. Simpson seemed to get as much pleasure out of teaching, as West did in learning. Later on, West would be the one to represent his ship in boxing or martial arts matches when they had competitions in port.

There were others among the crew that West had learned it was best to leave alone. There was talk around the ship that two of the men had been in a penal colony in Australia for several years. For the most part they kept to themselves. West found them to be quite surly and unfriendly. Occasionally they would play cards with the others but that was about all the "socializing" they participated in. It was rumored about the ship that one of them had been a boxer.

> My grandfather was a short man but was very strong even in his later years. He told me one time how his strong hands helped him out of a tough situation at a tavern. He was there with some of his shipmates having a good time when a group of locals decided to jump them. The biggest fellow was coming for my grandfather when my grandfather stiffened his forefinger and buried it into the fellow's throat at the bottom right above the sternum. Well the fellow fell down gasping and my grandfather started making a hasty retreat when a shot rang out and he felt his leg go limp. Someone had shot him in the calf but his friends grabbed him and whisked him out of the tavern to safety.

Chapter 4

Chapter 4

The First Shipwreck

After leaving the ports on the Baltic Sea, the Arabella headed up the coast of Norway. Norway was a beautiful, though very rugged country. The look of the country changed little through the years, even having the same architectural construction. The landscape was interesting with the many canals and fiords dotting the scenery. Norway is long and narrow, bounded by the sea to the north, south and west so it is not surprising that they have many skilled fishermen. The coastline in some ways is similar to Scotland. In viewing Norway from aboard ship, the one expression comes to mind is "peaceful."[4]

The Arabella stopped in several ports along the coast. Every time they went ashore into different

The First Shipwreck

countries, West picked up some of the language, even if they were only there for a few days. He had a prodigious gift for languages and he enjoyed being able to communicate with people in their own language.

When they had made all of their Ports of Call, the Arabella headed back from the Norwegian sea toward Scotland. The weather was miserable. Lashing rain and wind made the sea extremely rough, causing the waves to send huge sprays over the deck. All the men had on their slickers. Captain MacDonald was shouting orders, trying to manage to keep the ship stable. Suddenly the ropes on the mainsail snapped and the sail flopped dangerously. The men were huddled in groups trying to stay out of the path of the sail as it flopped erratically when the wind shifted in various directions.

There was a loud cracking noise as the main mast splintered and fell across the deck, ripping a gaping hole as it landed. This made the Arabella very vulnerable. When the waves washed over the deck, she began to take on water. Several of the seasoned seamen were seasick with all the lurching and rolling of the ship. After about an hour of this, the storm

The First Shipwreck

began to subside a little but the ship was listing badly.

Captain MacDonald shouted to the crew to find anything that would float and hang onto it. It became evident that they were in serious trouble. West had brought his canvas bag on deck earlier and had already wrapped it in some oilskin to protect it from the rain as it contained his precious books. Now, he took off his slicker and wrapped it around the oilskin bundle, tying it securely with ropes. It would be like a buoy!

Finally, the words they had been dreading to hear were shouted by the Captain, "***Abandon Ship!***" No one wanted to get into the frigid water. Some of the crew were crying and saying: "I can't swim!!" West had taken many a cold bath before but it was nothing to compare with this. Getting into the frigid water was breathtaking. He swam away from the Arabella pushing his buoy in front of him. Captain MacDonald urged everyone to get away from the ship lest they be pulled under when she sank. He also shouted to them to keep moving their limbs to keep the circulation going. The brave ones besides the Captain. . . Simpson, Gus, Alex, Jack, Harry, and West tried to get to the others and encourage them

The First Shipwreck

telling them to "hang on," "keep your chin up," "calm down," etc. but in spite of this constant encouragement, they were just too frightened. After about an hour and a half in the freezing water, Antonio, the cook, panicked and finally just gave up and slipped away. Another hour passed and three others gave up, including one of the former criminals.

The sailors had been in the water about two and a half hours when they heard a ship's horn! They could see a tanker coming toward them. That brought new life to the group! They began shouting and waving as much as they could with their teeth chattering and holding on to their floats. The men on the tanker threw out rope ladders and one by one they hauled the brave sailors out of the freezing water. Within a half hour of sighting the tanker, they were aboard a dirty, rusty, old tramp steamer tanker, but to them it was beautiful! They were wrapped in blankets and given some hot tea to drink. West could hardly hold his cup as his hands and fingers were so cramped from holding on to the rope on his precious "life saver" buoy. This had to be the best cup of tea he had ever tasted.

When they warmed up a little and got over the

The First Shipwreck

shock of almost losing their lives, West reflected on how sad it must be for Captain MacDonald to have lost the ship he loved so much and was so proud of and took so much pride in sailing her. A wave of nostalgia swept over West as he was thinking of another shipwreck that took place in the North Sea when he was about 3 years old. His parents were fishing in the North Sea and died in a shipwreck in a similar storm to the one he had just experienced. Thus, he had become an orphan.

The tramp was headed to the Belfast Shipyard for repairs but first they had to stop off at Glasgow to unload some barrels of oil. The repairs could be made at Glasgow but the captain of the tanker said he needed to unload the rest of the oil at Belfast anyway. All the hands went ashore at Glasgow. Everyone was glad to get off the ship for several reasons. It was very crowded with two crews, in addition, they were anxious to have some good food, warm baths, and new clothes! The men coming from the Arabella only had the clothes on their backs. The taverns were always glad to have the ships stay in harbor for several days so the men could spend their money on food and drink and girls! West had been saving his

The First Shipwreck

money so he was very careful how he spent it.

West decided he would be brave and visit his septuagenarian uncle since the tanker was to be in Glasgow about four or five days. He got himself all cleaned up with a shave, haircut and new clothes and felt like a new man, though he was only nearing nineteen! He went with the other hands to the tavern, "The Blue Lantern." He ordered the best steak they had. It covered the whole plate so the potatoes had to have a plate of their own. Food had never tasted this good. After the good nights' sleep, the new clothes and all this food, the men were in good spirits. They were laughing and telling the girls all about their shipwreck and narrow escape. Their survival was like a badge of honor. The real story was so exciting and scary that they didn't even need to embellish their tales like they usually did!

Chapter 5

The Return

West said his good byes and told Captain MacDonald and Simpson he would be back in plenty of time to load up with them, but in case he could not make it in time he would catch up with them in Belfast. He tucked the new wool shirt he had bought for his uncle into his canvas bag and set out for the Highlands. He was nervous and a bit apprehensive about how he would be received. Hopefully, he would be able to catch a ride with someone in a wagon, for it was a very long walk. He walked quickly on the cobblestone road among the gray, dingy, buildings near the wharf and proceeded into the city proper where the stone buildings were also gray, except for the green moss growing on some that were in the shade. Glasgow was a major port in Scotland with a number of industries near the docks. They had linen mills, factories, iron and steel industries. wool, silk, and other textiles and of course, their famous whiskey distilleries. The streets were deserted except for a few delivery wagons which were out early. There was a heavy fog that made the glow of the few lamps look surreal.

The almost seven years since he had left his uncle's place, had matured him significantly. He had become very strong and self sufficient, learning new things every day. He read every book he could get his hands on including the Bible and all the classics. At the age of 19, he had already experienced things that most people would not encounter in a lifetime.

He walked briskly, enjoying the cold moist air on his cheeks. His cap and muffler made him feel warm and snug along with his wool jacket and slicker.

The Return

The long walk ahead would certainly keep him warm.

Suddenly, the hairs on the back of his neck stood up and he felt a chill. He thought he heard someone walking behind him. His own boots made a noise on the cobblestones. He walked a bit further, then stopped. The other footsteps also stopped. He moved on, so did the other. He hastened his steps and turned the corner. When he came to a space between two buildings he stepped in quickly. He was trying to breathe normally but it was impossible. He set his bag down softly and waited. He heard the steps drawing closer, then all of a sudden he was face to face with the stranger.

"Why are you following me?" West said as he grabbed the front of the man's jacket. The stranger replied: "I've been paid to do you in, you cocky little bastard!" Out of the corner of his eye West saw a knife flash very close to his face. Just as he dodged, the blade caught him on the cheek. The man was taller than he but not as powerfully built. West was quick on his feet and called on all the fighting skills and defensive jujitsu moves that Simpson had taught him. He never thought he would be utilizing these skills to save his life! He had eagerly learned them just to compete in matches aboard ship and with sailors from other ships.

This time it was "for real." This man was trying to kill him! The blood from the cut on his cheek was running down his face but he didn't notice. He was too busy defending himself. After a few exchanges of blows by him and lunges by the man with the knife, West realized he had to do more than just be defensive. He had to attack! This was getting serious. He saw his chance and landed a karate chop on the man's forearm, causing him to cry out with pain and drop the knife. The thug quickly recovered and when he stooped to pick up his knife West landed a chop on the back of his neck. There was a sickening crack and the man fell motionless to the ground with a thud. West was sure he had broken the man's neck. West leaned breathlessly against the building trying to think what he should do.

Who would believe his story? He couldn't go to his uncle's now. If he told his uncle this unbelievable story, he would just say: "Ach, I knew you'd come to no good!" He might even turn him in to the authorities and what proof would he have that this stranger was the aggressor?!

The cut on his cheek was beginning to sting. He pulled the man further into the small space between the two buildings. He picked up the knife, wiped the blood off on the man's clothing and stuck the knife down into his own boot inside his pant's leg. He bent down and picked up his bag and as he leaned over he felt a little faint. He felt a rush of blood draining from his head due to the excitement and exertion of the fight in addition to the fear he had experienced. West gently straightened himself and brushed aside any thought of rest for he planned to be far away from this spot when the man was discovered.

West remembered seeing a public water pump just a little ways back, so he went there quickly to wash his face in the cold water. The water stung his cheek. He pulled a kerchief out of his bag and dampened it with the cold water. He held it on the cut for a few minutes then as the bleeding seemed to stop, he rinsed the kerchief and wiped the blood from his slicker. He was glad to have put on the slicker as it protected his new clothing!

West was grateful for the heavy fog as he walked to the edge of the city. He couldn't see much but then neither could anyone else! After he had walked a few miles he saw a haystack that seemed to be set off by itself. He burrowed into the side of it for he needed to rest. His heart was still pounding from the exertion and the fright of his narrow escape. Soon he began to relax and drifted off to sleep. When he awoke it was late afternoon. He was grateful it was still foggy but at least he could see that he was at the edge of an apple orchard. He crawled out of the haystack and helped himself to three apples. He ate one and put the other two in his bag. The apple was good but he was still hungry! Good job he had eaten that wonderful steak the night before.

  Fortunately, the cut was not too deep as it was on the thin skin of his cheek bone. He touched it gingerly and was pleased that a scab had formed. He had dodged just in time! A moment later and West would have been the one lying between the buildings!

  After walking around a bit, he saw a glimmer of light glowing through the fog. He walked toward it and spied a little tearoom, so he headed for it. There was a sign in the window saying "Ruby's Roost." The room was warm and cheerful, with wonderful smells coming from the kitchen. A smiling, well-rounded lady whom he assumed to be "Ruby," came to wait on

The Return

him. She brought cheese and scones and hot coffee which he devoured quickly and asked for more. This seemed to please her. She was very chatty but asked no questions, which was good. He sat there for a while reading an old newspaper and savoring the good food. He noticed that it was beginning to get dark, so he paid the lady and bragged on her cooking profusely. He wanted to get settled into his "haystack" before dark as he might not be able to find it in the fog. West slept well, wrapping himself in the woolen shirt he had bought for his uncle. He had certainly slept in worse places! He was up early as he knew he would not be able to return by the way he had come. He certainly wanted to avoid the place where the fight took place. There was a water pump close to the little tearoom so he headed that way. He glanced around as he leaned over the pump. There were a few people about but none paid any attention to him. The water was quite cold as it came out of the pump, but it was refreshing. He drank some, then splashed some on his face which he quickly dried.

From the water pump, he took a circuitous route back to his shipmates. While making his way back to the "Blue Lantern," he began to reflect on the

The Return

events of the past two days. Why would anyone want to kill him? He could not recall any major conflicts with anyone. His associations had mainly been with crew members. The men on the tramp didn't even know him well enough to hate him! The only "possible" suspect he could think of was the surviving ex-convict, Carl, who once jeered at him, calling him the "Captain's pet." West had just ignored his insinuation. He also recalled that he had beaten Carl once in a ship's boxing match . . . but to have him killed? That didn't make sense!

When West got back to the "Blue Lantern," he told the men that he had decided the distance was too far, so he had changed his mind. Someone asked about the cut on his cheek and West told him he got it "climbing through a fence" Later he told only Simpson the true story. He was glad to be back with his shipmates, but he experienced a certain sadness as the full realization came to him that he would not ever be able to go back to his beloved Scotland again. . **.and he never did!!**

The Captain of the tramp asked the crew of the Arabella to stay on as most of his men were staying in Glasgow. Captain MacDonald put it to the men to see if they were willing to do so and they all agreed. They felt it was the least they could do to help out the Captain of the tanker since he had saved their lives! The crew would have some time off while repairs were being made in Belfast.

    The trip to Belfast was a short one. The shipyard estimated the repairs would take as much as two or three months. The crew wasn't upset in the least by that as they all had friends or relatives to visit in Ireland. West had other plans. He asked Captain MacDonald and the Captain of the tanker, Edward Mahaney, if they would collect him in Liverpool when the repairs on the tanker were completed. He told them he wanted to go to a trade school and learn to be a machinist. The two Captains looked at each other wondering how West knew about this school but they didn't ask and he didn't offer an explanation. Seamen had an unspoken rule about not asking too many personal questions. The truth was West had seen an advertisement about a Trade School in Liverpool when he read the paper in "Ruby's Roost" in Glasgow. The Captains complied with his wishes and teased him unmercifully saying: "Come on West, are you sure it isn't a girl you're going to see?" West just smiled.

Chapter 6

The School

West was off on a ferry on his way to Liverpool the same day they arrived in Belfast. He located the Trade School and decided to enroll. He didn't have much money but he thought he had enough to last until his ship would collect him. If not, he could always get a side job. He wanted to be a machinist. He was eager to learn and anxious to begin the classes. He was fortunate that classes were just beginning. He met one of the other new students, who also was interested in becoming a machinist. They immediately struck up a friendship and their personalities clicked. He invited West to room with him at a boarding house just down the street from the school. His roommate was James Johnson from

The School

Wales. West told him his mother was Welsh-Scot so they felt a "kinship." They got along well (in spite of West's hair-trigger temper) and were both interested in their studies. James had more schooling than West and helped him with his math, whereas West had always been an avid reader and could explain the meaning of the texts. They were a good team. The time passed quickly because they were so involved with their studies. West had a keen mind and a fantastic memory. He remembered everything he ever read or learned. Because of his eagerness to learn, his mind became a sponge. This made up for his lack of formal education.

The school was not exclusively dedicated to book learning. The students did have some extra-curricular activities such as soccer, martial arts and rugby. The other students could not believe how proficient West was in the area of martial arts! He always won his matches no matter who he fought. The other "extra curricular" event was even more exciting. It happened about twice a week. There were several girls from a school nearby who would come over to watch their matches! Afterwards, they would get to visit with them in a visitation parlor!

The School

The weeks went by so quickly as the young men were totally engrossed in their studies and the school routine. This was the fulfillment of a dream West had yearned for during the years he had been at sea. He enjoyed the mental challenge the school provided him. All too soon, the day came when he had to leave.

West was a bit sad to leave when Captain MacDonald came for him, but he had learned a lot. He vowed to return when he could save enough money. His money had run out so he would have had to quit anyway. James was sorry to see West leave as he had been a lot of fun besides being a helpful partner in his studies. For West, who had few close relationships in his young life, this was a sad day.

One thing West acquired while he was in Liverpool was a seaman's box. Most of the deep-water sailors had one. It was almost a pre-requisite! This box belonged to a sailor in His Majesty's Navy out of Westhardlepool,

> *Every sailor had "a box" to store all their important papers, photos, and special treasures. Read the story "The Box" on page 121.*

The School

England. The Scotsman bought it for one shilling after about a half-hour's haggling! One shilling isn't much but it was all he had. It was a nice wooden box with a tray inside for special items. It would hold quite a few pictures of girls he had met and any important papers he had.

West collected his few belongings together along with a few extra books he had acquired at the school and placed them in his canvas bag and the seaman's box. He said "good bye" to James and the other friends he had made at the school.

He headed toward the docks with the box under his arm and Captain MacDonald by his side. Neither of them looked forward to going back aboard the tramp but still, at least it was a job. After the nice, clean, comfortable, quiet Arabella, the rusty, dirty, noisy, crowded, tramp steamer tanker was quite a contrast. They decided they would make the best of it until something better came along.

Captain MacDonald commended West on his studies and congratulated him for his desire to better himself mentally while he had the opportunity. This meant a lot to West and was a great encouragement for him. He had not received this much praise in his whole life. The usual remarks he was accustomed to were criticism, rebuffs and negative comments- especially from his uncle!

They made their way back to the wharf where the tanker was docked. The crew seemed glad to see West. When he told them what he had been doing, they were impressed. They said: "Now when something breaks, we won't have to go to the shipyard to get it fixed!" "Now hold on a minute" West said, "I only went to school for three months. I have to go back and finish when I get a few quid saved."

The tanker slipped away from the docks easily and it seemed to West that from the less noisy sounds coming from the engine room, the Belfast men had done their job well!

The School

When he was a young midshipman learning how a sailing ship works an old man was taking him around and "showing him the ropes." Which means quite literally that he was showing him which ropes controlled which sails and how to use them. Also, he was learning how to go up the ratlines to work the sails standing on the ropes that hang from the spars. But the old man told him that the most dangerous place to work was out on the bowsprit working the jibs and the flying jibs. This is because this area has the most movement, not only up and down but also side to side and in circles. The old man said "Boy, when you are out here on the bowsprit remember that it is one hand for me and one hand for the company. And when the sea is very rough it is two hands for me and forget the company!"

Chapter 7

Chapter 7

Another Ship Lost Another Long Walk

At least the tanker wasn't as crowded now since there was only one crew aboard! West cooked part of the time due to the tragic loss of Antonio during the shipwreck of the Arabella. It was a strange combination of jobs for West to be in the galley part-time and in the engine room the rest of the time. The men liked his cooking so much that they wanted him to do it "fulltime!" West *missed* Antonio terribly – not just for his cooking but for the companionship and conversational interchange they enjoyed daily. The water was calm so the tanker chugged along making good time.

The ship made stops in France and Amsterdam

Another Ship Lost, Another Long Walk

then to London. While they were in London, King Edward died. The year was 1910. The crew was staying in a hotel enjoying every minute of the ambience and good food when they heard that the funeral processional was to come right down their street. The street became lined with onlookers. The sailors climbed up the outdoor fire escape to the roof of the hotel for a better view. When the processional passed in front of their hotel, they couldn't believe what they were seeing. Royalty from many countries all over Europe and the British Commonwealth all dressed in their finery, passed by on horseback. Royalty dressed in velvet with their gold braided cloaks flowing over the horse's rumps. Even the horses had special plumes in their bridals. West counted sixteen kings on horseback plus numerous princes and nobles. They heard it said that all the "Crown Heads" of Europe were present! The hearse carriage carrying the body of the king was very ornate and was surrounded by men dressed in velvet and wearing long stockings. It was a sight they would talk about for months and an event they never forgot!

  The next morning when the crew was walking back to the ship, they were still talking about the

Another Ship Lost, Another Long Walk

funeral processional. As they approached within about one hundred yards of the dock, they saw a man stepping off the gangplank of their ship and under his arm he had West's box. When the thief saw the crew he started to run. In a flash, West took off after him. The man was not as fast as West so he dropped the box and ran away. West picked up  the box and examined it. It was still locked and undamaged. It was a good thing the man got away, for with West's quick temper and fighting skills the man wouldn't have had a chance!

  The tramp's engine continued running well after the overhaul they did in Belfast. After a full day of getting the cargo loaded, they were ready for the long trip. They had cargo going to the Hague, Netherlands, La Harve and Bordeaux, France, Lisbon, Portugal, and La Coruña, Spain. Normally, a tanker wouldn't carry this type of cargo, but since the tanker was available, they made use of it!

  All these Ports of Call were so interesting to

Another Ship Lost, Another Long Walk

West since he was the only one of the crew who could speak all the different languages. The Captain always used him to communicate with the longshoremen concerning loading and unloading of goods. It was comforting to be able to have someone on board who knew what was being said, lest there be any foul play going on!

It was always nice for West to go ashore and eat someone else's cooking. Some of the crew loaded up on the French pastries. The weather was nice and sunny, but not too hot. They would have liked to stay longer but the Captain said they needed to keep moving while the weather was pleasant. Portugal and Spain have always been popular vacation spots for the Europeans as well as the British Isles because of the warm climates of both. The tanker just stayed one night in each place.

To the disappointment of the crew, the tanker was headed to the Strait of Tangier or Strait of Gibralter by early morning. Gibraltar is a small peninsula extending from the tip of Spain. The Strait is between Spain and Morocco, joining the Atlantic Ocean and the Mediterranean Sea. They headed for Turkey with stops in Sicily and Athens. The sailors

Another Ship Lost, Another Long Walk

were always happy to go ashore in the Mediterranean countries, especially Italy and Greece because the food was so delicious. They loved bringing fresh vegetables aboard. It may seem like that food was all they thought about and that isn't exactly true, although, on long trips it gets pretty monotonous when you run out of fresh fruits and vegetables and have no way to obtain them.

They headed for Turkey with stops in Sicily and Athens. They spent a little time in Sicily eating wonderful Italian food, and purchasing a variety of pasta, some fresh vegetables, spices, and olive oil to cook with on board. In Athens, after the cargo destined for Greece was unloaded the crew took some time to visit the sights such as: the magnificent Parthenon, that was erected during the 5th Century B.C., and some of the Greek Isles. West was especially interested in the places and events he had read about in the Bible. The Apostle Paul went to Athens on his second missionary journey and spoke in the Areopagus. In the first century, Athens was considered the center of the intellectual world. After they had walked around and seen as much as they could, they headed back toward the ship. Someone

Another Ship Lost, Another Long Walk

told them there was another war in the Balkans. Word travels quickly around the docks, as sailors are anxious to learn about events that occurred while they were at sea. They all shared their information. West decided to put his box in storage for safekeeping, not knowing what was in store for the future or just when he would return to pick it up. He certainly didn't feel good about carrying it around in a place like Constantinople (now called Istanbul) which would be their next stop. He would retrieve it at a more convenient time.

    The trip from Athens to Turkey was uneventful until just before they reached the inlet where the docks were located. An Italian ship plowed into the tanker and it caught fire. The tanker was carrying oil so it blazed up quickly. Most of the crew were jumping off into the water, some of which was also in flames. They were frantically swimming away from the flames. Another ship nearby was successful in picking up most of West's group as they had been on deck ready to disembark. The crew was able to grab a few of their possessions before leaving the burning ship. West was glad he had put his box in storage! He managed to get his canvas bag! They boarded the

Another Ship Lost, Another Long Walk

other ship safely and at a safe distance, watched the tanker burn. It was fortunate that they had already delivered most of their cargo, but the articles they had loaded along the way went down in flames with the ship. As much as West and the others loved the sea, they were glad to step off on the dock after this harrowing experience!

They met some Turkish sailors who wanted them to stay awhile. Most of the crew didn't want to do this but there were no ships in the harbor. They were stranded. Constantinople was a very dirty city. They stayed six weeks, then they were asked to leave. The men were treated well for the first two or three weeks, after that things went down hill. They were having fights and arguments – it was time to leave. There was just one problem. They did not have a ship! The two Captains who had been staying in other quarters, decided to stay and wait for a ship. The other six men made the decision to walk back! It took them four months to walk through Europe to the North Sea. The group consisted of: West, Gus, Alex, Jack, Harry, and Simpson.

The "six" got together in a park across the street from their rooms where they could speak freely.

Another Ship Lost, Another Long Walk

They wanted to make their plans for the trip. They had many "discussions" on what would be the best route! One thing they all agreed upon was to skirt the Balkans as much as possible. This was necessary due to the war and "unrest" that was taking place. They wished they could circumvent them altogether but that was not possible. The basic decision of the route they would try to follow would be: crossing over the southernmost part of Bulgaria into Yugoslavia and Albania. Then they would follow the coast of the Adriatic Sea up to Trieste into Italy, Liechtenstein, Germany, Belgium, and finally, to Amsterdam!

They had a variety of experiences during those four months. None of the sailors had much money so they had to make it stretch. Sometimes they ate fruit they "appropriated" from orchards along the way to go with the cheese and bread they had bought. West showed them how to eat a "salt sandwich" and drink a lot of water so as to feel full. He had learned this from some Russian friends of his father, but he didn't mention that.

It was an ambitious undertaking but quite an adventure. Most people all along the way were very kind to them once they knew they had lost their ship.

Another Ship Lost, Another Long Walk

At first, people would be very suspicious, seeing this group coming along the road since they weren't exactly "well groomed!" After their story came out, people would offer them food and sometimes a place to sleep for a night. Even if it was a haystack or a loft in a barn, it was welcomed. At least it was dry and safe. More than once they had encountered some very unsavory characters who were bent on profiting from

> I had an experience when I was a kid of being hung by my belt on a bolt on top of a swing set. This prompted my grandfather to tell me that when he was a young midshipman his belt saved his life. He was working on a sailing ship up in the topsails and slipped and was falling to the deck when his belt caught on something on one of the spars that kept him from falling all the way down.

the "exchange." However, these thugs soon learned they picked on the wrong group! This group of seasoned sailors on first glance seemed to be in juxtaposition without even moving a muscle or saying a word. This in itself warded off many a conflict.

All in all, with what food they were able to buy,

Another Ship Lost, Another Long Walk

what people gave them, and what they "picked up" along the way, they managed an adequate subsistence! In spite of the facts that their shoes were completely worn out and what few items of clothing they had were in tatters by the time the trip ended, the men were in good spirits. It seemed as though they had faced the challenge of a lifetime and they had won! They had endured the extreme elements – the intense heat, the freezing cold, the rain, the snow, the wind, and to a man they never gave up. Most of them lost at least 20 pounds.

Chapter 8

Chapter 8

The African Adventure

The sailors felt like they were "home" when they arrived in Amsterdam. All the men had been in and out this port many times. In fact, West was born on a French ship right here in this Dutch harbour! His father was a Russian Jew and his mother was Welsh-Scottish. He mentioned none of this to his mates. They would ask too many questions! They would say: "How is it you have the name West? That's no Russian name!"

The grandest part of arriving in Amsterdam, besides the good food, was to have a hot bath, a shave, a haircut, and some new clean clothes! They walked in as a disheveled, unkempt, impecunious, group of "nair-do-wells." Now, to look at them sitting around a table playing cards, they looked utterly respectable! In fact, several could be called handsome! Sleeping on real beds, on clean sheets did a lot to restore energy to their bodies and to their poor worn and tired feet.

Amsterdam was an ageless city whose architecture has been maintained unaltered for centuries. It was one of the chief trading ports of call on the continent, usually with many ships coming and going. Every day the crew would check in at the docks to see if any ships were needing a crew. Times were hard in 1913. No one had money, it seemed. They were thinking that something had better come up soon because they were running out of money! They had pooled their "resources" after everyone had purchased their new clothes. Finally, word came to them that a Dutch ship was heading out for Africa and needed a crew. West always liked the Dutch ships because they were kept clean and had good food. As it turned out, those were the only good things about the trip! It was a very long, hard trip. It was this captain's first trip to Africa and they almost got lost. The Atlantic Ocean was rough and for the most part of the trip, the weather was extremely hot. They had too many ports of call. In most of the ports the

The African Adventure

stevedores had smiling, friendly faces and would sing all the time they were unloading the goods. On the other extreme, there were several ports whose workers were quite portentous and the crew was glad to leave before any trouble broke out.

Their stopover at Cape Town, however, was pleasant. The people were very friendly and seemed to be happy people. The Cape colony was one of England's possessions. West had read about the accomplishments of Livingston in exploring the interior of Africa, discovering the source of the Nile River in 1863.[5] Little was known about the interior before Livingston and others made their discoveries. Stanley's work on the Congo River brought about significant changes in what had been called "the Dark Continent." Trade in the interior, discovery of diamond and gold mines, and building of railroads, brought about significant changes, especially in South Africa. Johannesburg was beginning to become a preeminent city though as in most places in the world at that time, they had their distinct "class structure" as well as racial discrimination.

During this very long trip around the continent of Africa, West had plenty of time to read. He had already read completely through the Bible the year before so he now began memorizing chapters he especially liked and became interested in seeing how many he could quote from memory. He kept his mind active. He would get Simpson to read some verse and West would tell him which book in the Bible it came from and which chapter. West's interest in the Bible, no doubt came from someone in his early life who had a deep faith and expressed it to him at an early age.

As they were rounding the Cape, West thought that all in all, South Africa was an up and coming country and he would like to come back to visit at another time. It took several days to unload the goods in Durbin and even longer to load the outgoing goods. Once this was done, they moved up the coastline where they made brief stops at two other British possessions, Zanzibar and Nairobi. Somewhere along this coast they left the Indian Ocean where it joins the Arabian Sea, then into the Gulf of Aden and on into the Red Sea. The crew was beginning to get excited as most of them had not been here before. They were looking forward to seeing the Pyramids. West had told them about Egypt and read them the historical account from Exodus. They docked at Port Said and

The African Adventure

went over to Cairo for supplies and to look around. When they saw the Pyramids they couldn't believe their eyes. West wanted to walk around one just so he could see the immensity of it. Alex, Jack, and Harry decided that they would go with him but when they got up close enough to see how far it was, they decided they would just wait for him. They had all the walking they ever wanted on the walk across Europe! West went on his own and it was worth it. He couldn't imagine how the Egyptians had moved those huge stones in such perfect formations without modern machinery. After spending the day off the ship they were ready to get back.

The next day they were to go through the Suez Canal. West told them all about the Suez, how it joins the Mediterranean and Red Seas. It extends about 103 miles through marshy and desert land. It shortened the traveling distance between England and India by 6,000 miles. It is twice as long as the Panama Canal but cost 1/3 as much since it required less digging and has no locks. Hundreds of years before the time of Christ, canals were dug to connect the Nile River with the Red Sea. In A.D. 700 the idea

The African Adventure

was considered to join the Red Sea to the Mediterranean, but it wasn't until 1859 that construction was actually begun. It took 10 years to complete it! The Suez Canal has generated many conflicts throughout history and even to this day. This was probably more than his shipmates wanted to know about the Suez Canal as most of them weren't as interested in history as West was.[6]

The African Adventure

He told me that he and his shipmates got into an argument when they were on a merchant ship out of Liverpool in the Mediterranean about when they would make it back to England. Half of the guys thought in the near future and the others said it wouldn't be for a long while. Well, my grandfather was with the group that thought it would be soon and he was pretty sure of himself so his group said that they wouldn't shave or cut their hair until they were back in England. As luck would have it their ship got new orders and went through the Suez Canal into the Indian Ocean to make several voyages which meant they didn't make it back to England for over a year. Needless to say my grandfather and his friends were a scruffy looking lot when they docked. He said he was particularly scruffy looking because he had been burned in the face by steam some years back and couldn't grow a decent beard. It just came out in spots around his face. When he arrived at the boarding house he always used in Liverpool they wouldn't let him in because they couldn't recognize him and he looked so bad. He soon convinced them who he was and got settled. The first thing he wanted to do was go to the barber and get shaved and clipped. He told the barber to cut him close and shave all but the mustache. When the barber had finished he said to go ahead and take the mustache. Then he told the barber to just shave his head. Then to make the hair taking complete he told them to shave his eyebrows as well! Well, when he got back to the boarding house they wouldn't let him in because they couldn't recognize him again.

The African Adventure

The thing they did want to know is how much longer it would be until they got to Amsterdam! They got all their papers together as they had to be stamped to go through the Suez Canal into the Mediterranean. It took awhile as there were other ships ahead of them but when all the clearance was finished the whole crew perked up. They were on the last stretch of a very long trip. They only made one stop and that was in Athens, just long enough to pick up West's box that was in storage there at the docks. Next stop Amsterdam! They had been gone for four months.

After a few days' rest they were fortunate to get another ship right away. This time it was a fishing vessel. They fished the waters from the Irish Sea to Greenland and to Nova Scotia. They were fishing for herring and cod. West had forgotten how bad  the smell was on fishing boats! They were successful and caught plenty of fish but they decided they had enough of the smell. They sold the fish in Canada on the West Coast! The six men used several bars of soap and other toiletries to rid themselves of the smell of fish.

Chapter 9

Chapter 9

Two Wars

The sailors were on a roll now as they were able to get another ship called the SS Cloughton out of Westhardlepool, England. They boarded the ship in Halifax, Nova Scotia. It was a very good ship, clean and modern for the times. There were four Scots and the rest were Irish and English, plus Gus the Swede from the old Arabella group. Most of the Cloughton crew were in the British Naval Reserve, so they were more disciplined than some of the previous sailors West had sailed with, yet some had not had the experience that comes with being at sea for long months.

From Canada, the SS Cloughton headed back across the North Atlantic to England where they made

Two Wars

a brief stop in South Hampton. They were only there for two days and nights before getting underway toward Portugal. When they docked at Lisbon, they soon discovered that Portugal was in the midst of a revolution. This was not too uncommon for them. The Portuguese fleet was bombarding the fort. West and the other four Scots decided to go ashore and see what was happening.

When they got ashore, some men were running so they ran with them. The Portuguese rebels seemed happy to have the Scots with them. They had many guns so they gave some to the Scots. Someone set off a bomb near them and that got their attention! A group of soldiers began to charge the group but the sailors out ran the soldiers as they didn't want any part of what was going on there!

An elderly woman motioned to the sailors to come into her house. They were relieved to be out of the fray for a few minutes at least. The woman gave them some goat's milk to drink. The Scots gave her all the arms which included guns and sabers that the Portuguese rebels had thrust upon them. They told her she could sell them and keep the money. Using his language skills, West was able to communicate their thanks to her for her kindness to them. They quickly slipped out of the house into the dusk of the evening. The sailors skirted the city and climbed over several fences as they made their way back to the ship. It was dark as they climbed aboard and they went straight to their bunks as all was quiet aboard ship. They told only one man aboard ship about their adventure and that man was Simpson.

By sun-up the Cloughton was underway, headed for Buenos Aries. The trip was long and uneventful, though very pleasant. In the evenings after completing their chores the crew would play cards or read. They had a good crew who were very compatible. After spending a couple of days in Buenos Aries they continued on to Uruguay. The people in Uruguay were very friendly and wanted the crew to stay longer than just the time it took them to unload cargo and have a meal or two ashore. The captain reminded them they needed to move along as they still had a long trip ahead of them. They continued on through the Magellan Straits and stopped at ports in Chile and Peru. The crew was very pleased that they were able to be in Peru for about a week as it was such an interesting place. The awesome artifacts left by the Incas intrigued West and he asked many questions.

At each port, West would use his language skills to communicate with the locals and help the captain with any translations concerning the cargo and paperwork. The crew found West's language skills came in handy when they wanted to order food at the taverns or speak to some of the girls that were attracted to these foreign sailors!

About two months had passed since they left the Canadian coast when the Cloughton came alongside the Flamingos. There was another British ship nearby who was close enough to communicate with the men on the Cloughton. This time it was not vocal. The men on the other ship wig-wagged their message to the Cloughton. The message was this: "There is a war. The Germans attacked Belgium!" The year was 1914 and West was 26 years old. He had been to sea for fourteen years!

With this shocking news of the war, the Cloughton headed straight for San Francisco. Most of the crew on board belonged to the British Navy or Army Reserve so they needed to report for duty. When their ship arrived in San Francisco two of the

American reserve navy sailors on board left the Cloughton to join up with their mates who were on active duty. Before the ship could head out the remaining hands had to take care of the cargo and restock the ship. This took a while but finally they were on their way, headed down to the newly opened Panama Canal. They had to get in line with other ships that were anxious to get to their home ports as well. The Cloughton was the 96th ship to go through the canal. All the crew were excited to be a part of this moment in history It was a moment they never forgot! After they received their clearance to pass through the canal the whole crew sent up a cheer and they even forgot how impatient they had been while waiting their turn!

Their next "Port of Call" was Newport, Virginia. While the ship was being unloaded by the stevedores, West went ashore to make a few purchases. He entered a tobacco shop. In his heavy Scottish brogue he asked the storekeeper for "Twa pun o'bacca, sir." The man couldn't understand him and said in his Virginia drawl: "Wad youall say"? West looked on both sides of himself and saw no one else in the shop. West replied: "Errr ya speakin ta me?" He finally got the tobacco but he never forgot that humorous situation of his first encounter with the Southern drawl and the expression of "you all!"

The Cloughton was taking on some of His Magisty's Service Reserves by the time West got back to the ship. They were going back to England to join the war effort. Those on board who were not in the reserves had to go to other ships. West, Gus, and Simpson took a Belgian tanker which was leaving the same day.

They arrived in Amsterdam on Christmas Eve, 1914. Just as the tanker was entering the harbour a German submarine sent a torpedo hitting the stern of the tanker, causing a huge explosion and fire. Most of the crew went down with the ship. They were either killed by the blast or were wounded and drowned before anyone could rescue them. West, Gus and Simpson were spared as they were on the foreward bow. Another cold shock as the men hit the water with pieces of ice floating all around them.

West had his canvas bag that he had brought up on deck, preparing to go ashore. They did go ashore but not in the fashion they had planned! They had to swim a little ways and some longshoremen hauled them out of the water. The tanker was ablaze behind them. West was glad his "box" was still in storage!

Amsterdam had never looked as good as it did when the men came ashore after coming so close to death. As for West, he was reflecting on the past events and circumstances of his life when he had experienced near brushes with death. He thanked God for His mercy and protection!

> This is not a story but just a comment grandpa made one time. He said when he was in the British Merchant Marine during WWI that there was a great fear of the German U-Boats torpedoing the merchant ships and the ships going down quickly. In those days lifejackets were a new concept and most ships didn't have them for their crews. So when he was going about his daily duties he would carry a life preserver across his shoulder and keep his shoes unlaced. The shoes they wore then laced up high like combat boots and were hard to get off in a hurry. He said if it was necessary to abandon ship in a hurry, you wanted to get rid of shoes and excess clothes because they would pull you under in the water.

Chapter 10

Chapter 10

Smitten With Malaria And LOVE!

The three men, Gus, West and Simpson decided they would get away from this war. The only ship available was a banana boat headed for the West Indies. The three sailors signed on. At least there was no war going on in the West Indies and none of the three had ever been on a banana boat before! The trip to the Indies was fairly uneventful; no storms, no fights, and no illnesses.

Upon their arrival in the Indies, they soon discovered the climate was extremely hot and humid. They loaded the bananas quite green but due to the heat and humidity, before they arrived in the States to unload them, some of them gave off an acrid aroma

Smitten With Malaria and Love!

that was a close second to the fishing vessel – not as bad, but close!! West lost his interest in bananas as a favorite food, with the exception of banana pudding!

They didn't stay long in the West Indies as the "locals" didn't like them. There could have been trouble if they had lingered. Another motivation for leaving quickly came from a different direction. West contracted malaria and became quite ill. As soon as the stevedores got the boat loaded they headed for Galveston Island.

Galveston was a prominent port city just off the coast of Texas, in the Gulf of Mexico. The men remembered hearing about the 1900 storm when they were on the Arabella. The storm nearly swept Galveston away. Since that time Galveston had constructed a 13 foot seawall and had raised the whole city 17 feet.[7]

There were several reasons the men were glad to be going to Galveston. First, it wouldn't be such a long trip yet providing West a chance to recover from the malaria. Also, they had made friends with a tugboat Captain named Dick, who along with his partner Captain Daniels, could help them take care of West. Another reason that West in particular was

Smitten With Malaria and Love!

anxious to go to Galveston, was his interest in a 20 year old school teacher, named Bessie! West had become especially close to the two tugboat Captains and they had introduced Bessie to him. Once he had met her he couldn't get her out of his mind! She was so beautiful and sweet besides, NOTHING like the uncultured girls the sailors would meet in taverns.

> There was one instance where my grandfather was captain of a vessel. It was a banana boat hauling bananas from the plantation in Africa to the port. He told me one story that happened while they were docked at the plantation taking on a load of bananas. It was a particularly hot day and the ships hands decided to take a swim in the river. They had been swimming for a while and a small boy from a nearby village came by and told them to get out of the water because there were alligators there. The hands just laughed at him because they hadn't seen any alligators the whole time they had been there. They finished their swim and got on board and got ready to get under way. As the ship lurched upon pulling away from the dock, the ship's dog fell overboard and was immediately swallowed by a very large alligator!

Smitten With Malaria and Love!

West remembered little of the trip from the West Indies to Galveston as he was out of his head most of the time with a very high fever. He did remember hearing the men say they had docked in Galveston. Simpson was surprised that West was able to get out of his bunk on his own, as he had been in it for several days with the raging fever. He was a tough one. Once all the banana crates were unloaded, West, Simpson, and Gus went ashore. They got rooms at a small hotel near the docks. They all took baths, shaved and put on clean clothes. West promptly crawled in his nice clean bed. He gave Simpson some money and asked him to buy him some clothes, food, and quinine. Even though the weather was warm he was shaking like a leaf.

He tried to sleep and did doze for about an hour. When he awoke his sheets were wet with sweat caused by the fever. He had waked himself up screaming "swim, swim, you can make it!" This was just one of the reoccurring dreams he had. Another dream he had frequently throughout his life had to do with an incident that happened on one of his other voyages. It happened while West and some of his shipmates were sitting around a table playing cards.

The men had their shirts off since the temperature was quite high. West caught one of the men, Carl, cheating and told him to "get out of the game!" Carl left under angry protest and they went on with their game. West had his back to the doorway. After about 30 minutes Carl reappeared. Before anyone knew what he was up to, he flashed a big stainless steel fork and drove it into West's right shoulder! Simpson worked Carl over

> Another Fork Story. . .
>
> One day I was sitting with my grandfather on the porch of his house and it was a hot day and he was wearing shorts. He said come take a look and this scar on my leg. When I looked he pointed out three little scars that were dots all in a row on his thigh. He said that the story behind that was one day he was sitting at a sidewalk café with a young lady and they were getting along quite well. All of a sudden a big guy in a bad mood just came and sat down with them. He said "What are you doing here with my wife!" My grandfather said "I am sorry, I didn't know she was married!" Which apparently was not a satisfying answer because the man proceeded to pick up a fork and jammed it into my grandfather's thigh, grabbed his wife, and left.

unmercifully and took him to the brig. The others jumped up and got clean cloths and water to wash the blood off of the huge jagged wound. It eventually healed, but he was left with an ugly 4 by 2 inch scar. NEVER, EVER again did West sit with his back to a doorway even in his own home! Many times, years later he would awaken himself and others, dreaming and screaming reliving the incident.

Simpson returned with the food and clothing. West ate a little to keep his strength up but he really wasn't very hungry. He took the quinine. He thought he had gotten rid of this fever a few weeks ago but it kept coming back. He felt better after eating and got out of bed as he was unaccustomed to "lying around."

They all went ashore. It was good to be walking around on land again. They were surprised at how strong the wind was blowing. The year was 1915. West was 27 years old. It was difficult walking against the wind. West felt it more than the others as he was a bit weak from all the fever. Suddenly, it began to rain. It was a strong driving rain. They stepped into a doorway to get out of the rain. Some boys were running by and one of them shouted: "It's a hurricane. Seek cover!" The men look around and

Smitten With Malaria and Love!

Gus spied a big heavy truck parked nearby. The three sailors didn't waste any time diving under the truck. They clung to each other and braced themselves so they wouldn't be swept away. They wouldn't know until the storm was over that this 1915 hurricane was second only to the 1900 storm in intensity and lives lost. There were 275 people killed. The seawall and the raising of the city prevented a repetition of the horrors of the 1900 storm.

Gradually West's fever subsided and the attacks grew further apart. He was able to be up and about for longer periods of time. As part of his therapy in getting back his stamina, Gus, West, and Simpson would walk around the wharf and the Strand area. There were many ships in port so there was a lot of activity. Longshoremen were loading ships with various goods. There were bales of cotton stacked around everywhere. There were railroad cars with loads of sulphur and some with shells which were used for road building. One thing West always looked for was a tugboat owned by his friend, Captain Dick. He had become acquainted with Charlie Dick on previous trips to Galveston. More than once Dick's tug had guided ships West sailed on into the harbour.

Smitten With Malaria and Love!

They had developed a friendship over the years.

Sure enough, as West, Simpson, and Gus were walking along the pier heading for their favorite eating place, they spied Dick's tug. They motioned to him and invited him to join them for lunch. The four of them had a great time swapping sea stories and telling jokes in between bites of their delicious fried flounder and chips. Charlie invited them to come home with him and stay for the evening meal. West accepted but Simpson and Gus had other plans.

> West's forearms were 17", the same as the famous boxer Jack Dempsey's forearms.

Unknown to West, Charlie had also invited Bessie to join them so he had a pleasant surprise when she arrived. West was getting his appetite back after the malaria attack and it couldn't have come at a better time as Charlie's wife, Dema, was a wonderful cook. Dema was only a few years older than Bessie. She and Charlie had one child, Charlcie. Dema was also a fine seamstress. This was evident in the beautiful little dresses that Charlcie wore. Coincidentally, Dema belonged to a sewing club along with Ruby, Bessie's sister-in-law. Bessie knew how to cook and sew also since she majored in Home Economics in college at Sam Houston State Teacher's College in Huntsville, Texas. Her teaching career never left her much time to do the intricate type of sewing that the others pursued.

The evening went well. They had a lot of laughs with West telling of his adventures at sea. With his heavy Scottish brogue they would often ask him to repeat words so they could understand him. No other entertainment was necessary with West around! One of his favorite things was to hold himself up on his fingertips (with a fresh unboiled egg under each hand) with his feet straight up against a wall. Then he would lower himself down (still on his fingertips) to pick up a match off the floor with his mouth! He was so strong he was able to accomplish this feat even when he was much older. His fingers were like pegs!

Another betting situation . . .

It happened when he was in port and standing on deck with some of his shipmates watching a fight brewing with dockworkers on the dock. There were two large groups starting to mix it up and turning the dock into a big gang fight. My grandfather told his friends that he could stop the fight all by himself. They of course said he couldn't and he proceeded to show him he could. He went and found a large heavy block or pulley and tied it to a long cable. He then went down the ramp to the dock and started swinging the block around his head then started towards the gang fight. As he approached, the fighters could hear a loud swish-swish-swish as this crazy man was getting near them swinging this heavy gear that would surely kill anyone if they got hit by it. Everyone stopped fighting and made a path for him to walk all the way through the group and then he turned and walked back towards his ship. He said he started to get real worried that he was getting tired of swinging that thing and if he stopped those guys would kill him! He made it to the ramp of his ship, dropped the block, and scurried on board to safety.

Chapter 11

Chapter 11

Farewell To Captain MacDonald

World War I was underway in full swing by the time West went back to sea. Merchant ships were at greater risk than ever before, yet supplies had to get through!

West, Simpson, and Gus made a couple of trips between New York and the British coast with Captain MacDonald as their Skipper. It was good to be with this man again. He was a first-rate seaman as well as a first-rate leader of men. Not everyone had the talent of welding a group of individualistic, hardheaded men – with a few misfits thrown in – into a team of hard working, cheerful sailors. Captain MacDonald had that skill. He was tough, though he could be very gentle. He took his job seriously and wanted the work done correctly. The Captain also had a great sense of humor that he often used to dispel potentially explosive situations among the crew. Most of the men were very loyal to their Captain.

Of course West had a special devotion to Captain MacDonald due to the way he took West on board as a cabin boy at age twelve! He had treated West like a son, training him and encouraging him on occasion when he was lonely. Though he was just a lad and required some special attention and direction, West always tried to "pull his own weight." However, reflecting on his early days, he supposed it would be possible for someone as warped as Carl, to misconstrue the special "attention" or "instruction" West received from the Captain, as favoritism. Thus, from time to time Carl would taut West with the title of "Captain's Pet." That was all in the past. Carl was turned over to the authorities after the stabbing incident.

There were tense moments on these two voyages under Captain MacDonald. The German submarines were always on the minds of the crew. They had witnessed a couple of their sister ships being struck by torpedoes and it was not a pretty thing to watch. On one occasion they had been able to rescue some of the survivors. This act alone brought chills to the spines of the four men on this ship who had been part of those rescued by the tanker in the North Sea when the Arabella sank in a storm. By all odds, the four of them should have drowned! Also the experience in the Turkish harbour when they were torpedoed was still fresh on their minds.

At the end of the second trip, Captain MacDonald told the men this would be his last voyage. He was retiring to a little place in Scotland that he and his wife had purchased years ago. It was on the coast where they had a nice view of the sea and a nearby lighthouse. His wife, Annie, was looking forward to having him home year round instead of just a few weeks here and there between voyages.

The men all shook hands and the four hugged one another. It was a bittersweet farewell. They were happy for the Captain to be able to enjoy his later years with his wife, yet sad that they would probably never have the pleasure of seeing him again.

Chapter 12

Chapter 12

The Orphan

Captain MacDonald's leaving for home caused West to think of the lovely lady waiting for him in Galveston! West had never had a real home. His parents were shipwrecked in a storm while fishing in the North Sea. He was three years old at the time. Fortunately, he was not on the ship with them when the it went down. He was staying with his mother's sister, Margaret, and family. Margaret's husband was a minister of the Free Church of Scotland in a small village called Logie-Coldstone. This family had children of their own who doted on West who was a handsome child with dark curly hair and gray-green eyes that missed nothing! There was no one else to take him in. This connection is responsible for his keen interest in the Bible. Each evening the family would gather around the hearth where a cozy fire of coals were glowing, to hear the preacher read from the Bible and explain the significance of what they were hearing. Even the little three year old West would sit with rapt attention taking in every word, as this man was a great story teller. West himself became a good storyteller and entertained his own children and friends not only with Bible stories but often stories of his own adventures some 40-50 years later!

    West was a happy child and the other children helped their mother care for him, as she was not a strong person. They gave him the name of their own son who had died at an early age. As time went by Margaret's health began to fail and she became weaker and weaker. She developed pneumonia and died only a few years after West came to live with them. This presented a problem of what to do with West. The other children were able to look after themselves but West being about six needed more supervision. The decision was made that he should go to the home of his mother's brother who lived alone in the Highlands on a sheep ranch. Since West was a sturdy boy maybe he could be of some help to the

The Orphan

uncle.

As it turned out, this was a traumatic change for West! For him to leave this warm, loving family and move in with a cranky old bachelor who didn't even want him, was a big adjustment. He was expected to do the work of a man. He had to go out in the bitter cold to care for the sheep and do any repair work that the uncle needed. He also did most of the cooking as he had helped the cousins prepare food for the family. At least he knew how to do a few things. This was one job he didn't mind doing! He liked to make bread and learned many ways to prepare lamb! Occasionally, when it was warm enough he would go to the nearby stream and catch some fish which was a nice change in their menu. Life here wouldn't have been so bad if his uncle had not had such a disagreeable disposition. He would take a strap to West if he didn't do the work to suit him. It was on such a day when West was 12 years old that he made up his mind to run away, sign on a sailing vessel and make his own way.

Now, in his early 30's reflecting on his own childhood, he wondered what it would be like to have children of his own. Would he be a good father? If

The Orphan

Bessie would have him for a husband he felt sure she would know how to raise children. She had five brothers and one sister. From things she had told him about her family, they were very close. Her mother died when she was five so her brothers looked after her and her younger sister. Her father married again after a few years but the brothers were so used to caring for the girls they just continued doing so until they were grown!

Chapter 13

Chapter 13

Back To School

After the departure of Captain MacDonald, the group (West, Simpson, and Gus) made a few short trips – mostly bringing goods into England. They made one long trip to Australia and back. By the time they returned from Australia the war was over. There was dancing in the streets, singing, and, of course, crying for those whose loved ones would not "come marching home."

West made a decision. As long as he was here in England he would go back to Liverpool and finish his studies so he could receive his Machinist papers. If a man had that certificate no matter whether you were working on land or sea, you would receive higher wages!

Back to School

He said his "good byes" to his buddies and asked what their plans were. Gus said he was going home to Sweden for a few years but he might see West later in Galveston. Simpson's plans were to hang around the British Isles for a while and look up some old girl friends, if they weren't already married!

West headed for Liverpool and spent the next six months applying himself to his studies. It wasn't the same without James, but now that he had a girl in the States his mind could be focused strictly on his studies. He wasn't interested in the "extra curricular" events! He really enjoyed learning new things. Since he had more practical experience, all his studies were easier for him. Everything he read made more sense. The time passed quickly, with his Machinist Papers in his hands he was soon looking for a ship bound for the Gulf coast of Texas. There was a ship leaving in two weeks and needed a man in the engine room. They were waiting for some wounded American soldiers to get well enough to travel. West signed on and was pleased to show his newly received "Machinist" papers!

Seven years had gone by since West had first laid eyes on the lovely Bessie! He had seen her a time

Back to School

or two in the interim but only for a short visit as he was in and out of Galveston. Now he was missing her sorely. One unsettling thought crossed his mind. Maybe by now she had found someone else. Just the thought of this caused a sick feeling in his stomach.

He had saved a bit of money during the time since he had attended the trade school. This gave him a little more confidence in approaching her concerning marriage. Would she even consider marrying this "rough and tumble" seaman, and a foreigner to boot? What would her brothers think? West was now thirty four and Bessie was twenty seven. If he had any serious intensions he knew he should act on them soon.

The freighter was finally leaving Liverpool, a week sooner than expected and due to the illness of the engineer, West was able to sign on as "chief engineer!" What a break for West. The pay was good and it was a good sturdy ship. It was loaded with some cargo and quite a few Americans returning from the war. They, too, were anxious to get to the States.

There was a touch of Fall in the air. The weather was beginning to change. Not that it mattered to West as the Engine room was always hot! The crossing was uneventful, though it seemed long. But then, he had never before felt such an urgency. Most of the time he had been resigned with the attitude of "we'll get there when we get there." Now he had a reason to want the time to pass quickly!

Chapter 14

Chapter 14

A Life Changing Experience

As the ship approached the Gulf of Mexico, there was an unusual aura of excitement among the passengers as well as the crew. The crew was looking forward to a couple of weeks ashore while they were waiting for their cargo to arrive. They had enjoyed this trip. It was the first time in several years that they were able to relax a little and not be concerned about being torpedoed or bumping into a mine. The small group of passengers, who were all service men, were anxious to get back to their families. Three were Navy men and the other two were Army. All five had been injured in the heat of battle and had spent several

A Life Changing Experience

months recovering in hospitals in England. All five had experienced serious injuries and had just now become "fit for travel." The ship's crew tried to give them special attention, making the trip home as comfortable as possible.

One evening some of the crew even put on a little entertainment show for them, making them laugh for the first time in a year. One of the sailors was a juggler, one did some slight of hand tricks with cards. West did his "strength" trick. West's other exhibition required the assistance of Simpson. West had taken fencing lessons in London from the same instructor as Erol Flyn, when Flyn was preparing himself for the first edition of "Muttony on the Bounty," one of his early "swash-buckling" movies. West had become an excellent "fencer" and had taught Simpson some of his skills. The two of them put on quite a show and the men loved it! Among the cast of "entertainers" was an Irish tenor who was a big

> *In later life, to the chagrin of my mother, he performed this "strength" feat many times for guests in our home – including the occasion of his 80th birthday!*

A Life Changing Experience

hit with his medley of Irish songs. He also did a little jig that made everyone laugh. His job on the ship was "communications." He certainly communicated to the injured men that day!

After weeks at sea, the day finally arrived when the ship entered the Gulf of Mexico. As they came near to Galveston Island, West sighted the familiar outline of his friend's tugboat. In a matter of minutes Captain Dick's tug was along side of their ship guiding it into the dock. There was more excitement aboard the ship than usual. With the service men so anxious to greet their families for the first time in several years, not knowing what lay ahead for them, and West's anxiety concerning his relationship with Bessie, tensions were running high.

When the docking was completed, the gang plank put in place, the crew helped the injured service men to disembark into the arms of their waiting loved ones. There was a high school band playing, flags were waving and cameras flashing, welcoming the five valiant men back home! The crew was surprised, but pleased, that this much patriotism was exhibited since World War I had been over for more than a year.

The anxiety was needless! The families of the

A Life Changing Experience

wounded veterans were so very happy to have their men home again, regardless of their condition, that their fears were allayed the moment they were united with their loved ones. As for West, he had some chores to complete aboard ship before going ashore. He accomplished his tasks in record time when Captain Dick relayed a message to him that Bessie was waiting for him at the gangplank!

They were married June 23, 1923

Times were really hard during the depression. West took any job he could find. He worked as a plumber, then for an oil company, then finally he worked at a Dry Dock, a job he really liked because it dealt only with ships!

For a man who had never had a *real* home or had never known the warmth of a stable family, West did remarkably well. Of course he and Bessie had

A Life Changing Experience

their "ups and downs," but no more than any other normal family. There was never any doubt that they loved each other. He thought Bessie was beautiful and she thought he could do anything!

There were a few cultural differences, such as she wanted him to go around in a white shirt and tie like some of her 5 brothers did. His kind of work did not lend itself to that. Also, there were a few disagreements on how certain things should be cooked! Like the old adage says "too many cooks spoil the broth." The broth was never spoiled but it might have been stirred once too often. Bessie had been to college and studied "Home Economics" while West had cooked aboard ships of many nationalities. Bessie was a School Teacher and West was a Machinist. Though he was a "blue collar worker," he was self educated. He was an avid reader, reading all the Classics. He had read the Bible from Genesis to Revelation many times and could quote whole chapters from memory. He did not have a "sailor's" language and I never heard him take God's name in vain. Who would have dreamed this "runaway" would be so happy to settle down to "family life?!"

He was baptized when my sister was about five

years old. When he walked down the isle for the baptism, he was followed by his friend Captain Dick, the tug boat captain. West became an elder in the church in Galveston, as well as a Bible class teacher. His brogue plus his knowledge of the scriptures, made his classes of great interest for many years.

In 1944, while in college, Olga West, majored in Art and drew the church building where West was baptized. It was built in 1895. The West family attended the Church of Christ at this location for many years.

A Life Changing Experience

All of their differences faded away as the years went by. One thing they both agreed upon was their pride in their two daughters and we thought we had the best parents of all. After all there weren't any other dad's (that we knew of) who spoke eight languages and had sailed around the world three times! And our mother was creative and a beloved Elementary teacher. Upon retirement they both enjoyed doing the things they had no time for in their earlier years. West created many artistic flowers and also jewelry out of copper. He also helped friends with mechanical problems and repairs of all kinds.

West and Bessie were extremely hospitable the rest of their lives. During World War II many service men and medical students were invited to eat with us for Sunday Dinner. West was working at Todd Galveston Dry Docks as a Machinist Leaderman repairing damaged ships and continued this work all through WWII and several years after until retirement. They lived in Galveston in the same house for forty four years.

Evenings were a special time for West. After a good meal and the sun had gone down, we would sit on the porch enjoying the Gulf breezes. The stars and the moon were the big topics of our conversation. West would point out the Big Dipper and the Little Dipper and other constellations that sailors relied on for directions. West had his glass of hot tea with milk by his lawn chair – always a glass, <u>never</u> a cup!

Who would ever believe this crusty old seaman would leave the Gulf Coast?! He had always said he would never leave the water. After my sister, Olga, graduated from college she became a commercial artist. As a newlywed, she and her husband lived in Houston, Texas where they raised three daughters, all talented in their own way. When I got out of college and married, I spent all my married life in West Texas: First in Tulia, next in Snyder, then in Monahans, and finally in Lubbock. My husband was president of banks in each of these towns, except Tulia where he started his career and Snyder where he was Vice President.

> There are roads and roads that lead to Rome,
> But they don't lead down to the sea;
> And they take me not to my island - home
> So they're not the roads for me.
>
> Unknown author

A Life Changing Experience

After we had been in Monahans for several years, my parents (West & Bessie) decided to sell their house and move there! We were happy but could hardly believe it, for Monahans is almost in the desert, a very dry and desolate oil town. It was not even near a lake!

I would like to think my parents moved there to be with me. I soon discovered the real reason was their four grandchildren! One might think they would be lonely for their friends of 44 years and one of my mother's brothers and wife who lived in Galveston. I'm sure they missed all of them but they made new friends immediately by inviting people into their home for a meal. (And of course, storytelling and tricks by my dad!)

They had a happy life there. People were so intrigued by West's adventure stories that word got around town. He was interviewed on tape by a lady who was working on her master's degree. He was also interviewed by several other people including a reporter from the local newspaper. The

See the reprint of his interview with "The Monahans News" on page 125.

A Life Changing Experience

article came out on the front page along with his picture. When asked to tell some of his adventures, he would usually say something like: "You wouldn't believe me if I told you!" This was true. He had many narrow escapes and unbelievable rescues during his life aboard ships.

During this time in Monahans, West refurbished a large sailboat we had purchased at Lake Travis (near Austin). It was transported to Monahans on a cotton trailer and remained in their front yard for many months during the repair process. It was quite a conversation piece in this very dry West Texas town. People would drive by slowly with their children pointing to the boat. A very talented local artist who lived across the street from West and Bessie painted several pictures of the boat. It was a large wooden boat with a cabin and 2 bunk beds and a 30 foot mast. West scraped it, painted it and polished the brass portholes. He checked out all the fittings on the mast, replacing any broken parts. It was a big job but I think he really enjoyed the whole process even though it took months. Finally, the day came to transport it to a lake where we had a cabin, near Abilene. People who were driving down the highway are probably still

A Life Changing Experience

talking about what they saw.

It was a great boat and many teenagers got to ride on it. It would hold about 15-20 at a time. West got as much enjoyment watching them as they did on the ride.

After a few years passed we moved to Lubbock and were fortunate to find a house that had a guest house on the property. This was perfect for West and Bessie in their declining years. They were able to be near yet able to have their independence. West did all types of crafts with metal, mostly copper. He made flowers, flowerpot holders, jewelry, candle holders, etc. Bessie did the painting if any was required. They also enjoyed planting a garden, though by this time he was in his 80's and she was in her 70's. West even won a blue ribbon at the Fair from some of his products. I think he got a real kick out of competing with all the women.

One of the grandchildren told his English teacher about his grandfather's love of Rudyard Kipling's work and that his grandfather had a Scottish accent. She promptly invited West to come and read to the class. He was more than happy to do this and chose his favorite "Gunga Din." He read from his own

A Life Changing Experience

book and the children loved it, even though they couldn't understand some of it. He told them of the time he went into a restaurant in London and spotted Rudyard Kipling sitting at a table near by. West got up from his table and walked over to speak with him. He said: "Mr. Kipling, I have read all of your books and enjoyed them very much." Mr. Kipling replied: Good. Good, That's good. I am glad that you like them. Thanks for telling me."

West enjoyed using his language skills with foreign students. Since he could communicate in eight languages he was always finding someone to talk with in their own language. They in return were happy to find someone who knew their language. "This runaway" lived a full and useful life, always helping others until his death at the age of 84. He could have been much older. We were never quite sure of his age or his birthdate – neither was he. We always thought he just "picked one out." The same is true of his name. He used to say "he was born on a French ship,

A Life Changing Experience

in a Dutch harbor, his mother was Welsh-Scottish and his father was a Russian Jew." We have never been able to affirm or deny this statement.

Bessie died at the age of 86. She was 7 years younger than West. She possibly would have lived longer had she not been raped by an intruder when she was 81. That tragic event hastened her physical and mental decline.

Both West and Bessie left their mark on a large number of people, both young and old, whose lives were touched by their hospitality, warmth, and love.

"Nothing in the Past is dead to the man who would learn how the present came to be what it is . . ."[8]

The Box

by William G. West

This is a story of a box. Most deep-water sailors have one. This box belonged to a Welshman who was an A.B. on a ship of the line.

I was doing fine in the H.M.N. till one day I was carried ashore and put up for sale. After half an hour of haggling, I was sold to a Scotsman for one shilling. Now one shilling is not much; but, when that is all he had, it is a fortune.

So began my life on a dirty tramp steamer. It was always hurry, hurry from one place to another. We never stayed long. It was the year King Edward died. We saw a lot of kings, princes, and nobles on horseback. They had all lost their jobs.

We also had a few accidents. One time I almost burned up. But, we came out without a scratch. One time I was stolen, but the thief could not run very fast. He dropped me.

Well, we lost another ship. We did not like the rusty old tramp anyway. Now there was one more war in the Balkans. I was put in storage crammed full of pictures. It seems every girl had a picture to give away. Well, I heard the war was not much. It seems the Turks asked everyone from the ship to stay awhile in Constantinople, a very dirty city. After about six weeks they were told to leave. My owner and about five more decided to walk back. It took four months to get to the North Sea.

Then the hurry started again. Times were hard in 1913. No one had a job; no one had much money. We had to take a ship to Africa. It was a very hard trip and we almost got lost. We decided not to go to Africa again. We went deep sea fishing from the Irish Sea to Greenland to Nova Scotia. We fished for herring and cod. We got tired of the smell, so he decided to take me to the west coast of Canada.

We shipped out on the S.S. Cloughton out of Westhardlepool. It was a good ship. Four of us were Scotch and the rest were Irish and English. Most of the group were in the Navy Reserve. We came to Lisbon where they were having one of their revolutions. The fleet was bombarding the fort. We four decided to go ashore and have a look and here is my owner's story:

We saw some men running and we ran with them. They were very nice. Some had too many guns, so they gave us some. Then someone set off a bomb. Some soldiers began to charge us. We outran them and an old woman told us to come into her house. She gave us some goat milk. We gave her all the guns and sabers and told her to sell them. We climbed over some fences and eased our way back to the ship. We only told one man about the adventure.

Next, the ship took us to Buenos Aires and to Uruguay, then through the Magellan into Chile and Peru. After a time we came alongside the Flamingos. Another English ship wigwagged – "There is a war. The Germans attacked Belgium."

We came into San Francisco. Most of the crew belonged to the Navy or Army Reserve. So, most of them decided to go to war. But first we took care of the ship and the cargo. After some delay we sailed through the newly opened Panama Canal. Ours was the ninety-sixth ship going through. We left this ship at Newport and took a Belgian tanker. We arrived in Amsterdam on Christmas Eve in 1914. The tanker was sunk and I was put in storage.

My owner did not stay long. He lost another ship. He said the water was cold--very cold. Most of the people lost their lives. We left for the West Indies as the war was over for us. That was a mistake. The people did not like us and told us to leave immediately. I was left on a banana boat. After a while we came to Galveston. My owner decided to stay until he got rid of the fever he had caught somewhere. We left Galveston about five times and came back. So, we decided to stay. It was hard to stay on shore. Then, all at once something happened. A woman came into our lives. It was *the* woman. My owner turned me upside-down. All my pictures had to be burned, the girl pictures, that is. She is now my half owner.

I never did belong half and half. Well, one day she got curious and wanted to open me. She tried every key. That was funny. My lock was broken by a thief ten years earlier. So one day the man opened me with a pocket knife. That was fun. I was kicked under the bed. Once in a while I was hauled out and we remembered a few things. Now after fifty-seven years, I was given a cleaning, to be a present to a grandson. Now, Rodney, be good to me. I am old and have sailed the seven seas.

The Monahans News

Member Associated Press
Manahans, Texas, Thursday, May 4, 1967

Tales of The Sea

Retired Scotsman Recalls
Sailing Days
By Lucila Rogers

The taste of the cold North Sea still lingers in the memory of 79-year-old Scotsman, William G. West. "I sometimes think I can still taste, the salt water and feel the icy blackness."

His voice still carries a strong Scotch burr and he hasn't lost his strong sense of humor or quick spark of temper that sent him running away from school at age 12 and off to a sea on a sailing ship.

Today he makes his home in Monahans far from the shores of his homeland. With his wife

Bessie, he keeps busy at various hobbies of specialty cooking, copper jewelry making, and raising flowers. He became a U. S citizen in 1922 in Galveston, TX, his home for 44 years before moving to Monahans six years ago. Mrs. Willard Paine of Monahans and Mrs. C. K. Money of Houston are his two daughters. In addition, he has seven grandchildren "all of whom are 22-karat gold," said the proud grandfather.

His clear blue eyes crinkle with easy laughter but have a piercing look and he is not inclined to live in the past as is common with so many elder citizens. "Let the past lie," he said briskly and he refused to give his old Scottish name. He had to be coaxed really to reveal much of his past history, saying of such tales "you would not believe me if I told them to you anyway."

Be that as it may, the wry little man and ex-sailor did tell that he had been shipwrecked no less than three times. The first time left him an orphan at the tender age of five years. How else could he help but be a sailor? For he was even born on a ship. His father owned a sailing vessel – a three-mast schooner – and his mother was traveling with him at the time of the birth of William G. West. He says his birthplace was the "North Sea." Both parents perished in a storm that saw the young boy as the only survivor of the shipwreck. He was raised by an uncle until at the age of 12, the sea beckoned too strong and he ran away.

He says he has visited every major port in the world. For over 16 years he sailed the seven seas. Once, just for the fun of it, in Cairo, Egypt, he walked completely around the Pyramids.

"You really want to hear about some of my adventures?" he paused to ask, before going on to sketch briefly how one time he served on a gun-running ship smuggling arms to Greece, during the Balkan Wars of 1900 to 1906.

And of how during World War I, the war had been going on for 40 days before he and his ship's crew ever heard about it. There was no radio or wireless in those days and the Panama Canal wasn't open so they were busy making the long voyage around Cape Horn.

On Christmas Eve, 1914, he landed in Amsterdam, Holland. Right after he got off the boat, the ship was sabotaged and went up in smoke in the harbor. Later he was on a British ship in the North Sea when she hit a mine and all hands spent hours in the cold, cold water before being rescued. In another incident in the life of this adventure ridden sea faring. man, he was chased out of British Honduras and told to leave the country. He didn't relate just why . . .

West may have left schoolhouses and formal education behind, but this gentleman can speak eight languages. During World War II, he worked for the government in Galveston as a foreman train. In machinist and other shipyard operations for the War Production Association.

"I saw 16 kings on horseback!" he further related, This occurred in London when the state funeral of King Edward Tudor was being held and he and some companions rented a hotel room providing them with front-row seats for the funeral march.

"What'cha want hear all this for?" the old Scotchman kept asking. "It's all so long ago and far away . . ." But the tales spring easily to his lips once you get him started and no doubt, he has many more sealed up for later visiting. An interesting resident and a welcome addition to Monahans, but he says he thinks he is going to pack up and leave if it doesn't hurry up and rain.

The voyage from sea water to the dry lands of Went Texas has been long. But a man's memory can span the gap and bring the smells and sounds alive again.

Epilogue

In the course of researching information concerning my dad's life, my youngest son, Rex, and I had several serendipities while traveling in Scotland. The first occurred on the train as we traveled from London to Edinburgh. Rex was having a conversation with a business man across the isle from us. He was speaking rather loudly in order to be heard above the noise of the train. Rex was explaining the purpose of our trip.

A lovely young woman sitting a few rows back could tell we were from the United States, and became interested in our quest. I invited her to come and sit with us and we explained our search for information. We discovered that she was on her way to Balmoral Castle in the Scottish Highlands to look after the Queen. This young woman, Angela Kelly, was the Queen's Dresser! Since that time she has been promoted to Her Majesty's Personal Assistant and Senior Dresser. Angela was so very kind to us. She insisted that we take her lunch that had been prepared for her at Buckingham Palace. She knew we were planning to rent a car and that we would not have a chance to eat for several hours. We had an interesting visit discussing our "heritage." After a long and busy period, Angela and I are enjoying corresponding once again.

Another interesting event occurred in a tiny little village of Logi-Coldstone, in the Highlands of Scotland. Rex and I went into the only commercial building (that we could see). It was a little grocery store. I made some inquiries to no avail. At this same crossroad was a church building and a school.

I stepped into the school and was received graciously by the head Mistress. She took me into one of the classrooms and introduced me to the teacher. The teacher introduced me to the class and asked me to say a few words to them. I didn't have a clue what to say to them so I just began by asking them some questions such as: How many of you have visited the United States? Do you have any relatives living in the US? I went to a map on the wall and pointed to the part of Texas where I lived. I asked if any of them had been to Texas. I was surprised at all the hands that went up! When I asked if they had any relatives living in Texas, they all wanted to tell me about their relatives and their experiences while visiting them. Of course time did not permit this. I felt I had taken enough of their time, so I just thanked the teacher for allowing me to interrupt their class.

Rex and I went on our way. The other serendipity came when we were looking for a hotel or bed and breakfast to spend the night. We discovered the park area where the poet, Robert Burn's spent many hours writing his poetry. It was a beautiful, quiet area with a natural waterfall, making it a delightful place to rest. I had been to Scotland several times before but had not come to this location. We found a great bed and breakfast not far from there. We thoroughly enjoyed our whole trip and appreciated all the kindness extended to us everywhere we went. I hope to go back sometime.

Endnotes

1. Montgomery, D. H. "The Leading Facts of British History" circa 1901. Copyright 1887. p. 409.

2. Houston Chronicle. September 3, 2000. Special Report. "Echoes of the Storm."

3. Montgomery, D. H. "The Leading Facts of British History" circa 1901. Copyright 1887. p. 402.

4. Kleivan, Kare. "Norway's North Country."

5. Montgomery, D. H. "The Leading Facts of British History" circa 1901. Copyright 1887. p. 403.

6. Encyclopedia Britanica. Vol. 17. p. 767-768.

7. Houston Chronicle. September 3, 2000. Special Report. "Echoes of the Storm."

8. Stubb - Constitutional History of England. By D. H. Montgomery. Copyright 1887.

Antidotes provided by Rod Paine,
grandson of William George West.

Printed in the United States
36397LVS00005B/109-147